Table of Contents

1 Introduction to Energetic Materials

Energetic materials, a broad term encompassing compounds designed to release large amounts of energy rapidly, are crucial in various applications. From propellants driving rockets into space to explosives used in controlled demolitions, these materials significantly impact our world. Understanding their fundamental properties is essential, whether you are a researcher, engineer, or simply curious about their nature. Their behavior is governed by complex chemical reactions, making them a fascinating area of study.

The classification of energetic materials often hinges on their intended application. High explosives, for instance, detonate violently, releasing a tremendous amount of energy in a short time span, resulting in a powerful shock wave and intense heat. Propellants, on the other hand, burn relatively slowly and steadily, producing thrust rather than a destructive explosion. Pyrotechnics, a third category, generate light, heat, smoke, and sometimes sound through carefully controlled chemical reactions. Each type requires a unique combination of properties for effective functionality.

Historically, the development of energetic materials has been closely linked to advancements in chemistry and technology. Early explosives, like black powder, relied on simple mixtures of naturally occurring substances. The discovery of nitroglycerin in the 19th century revolutionized the field, paving the way for more potent explosives like dynamite. The subsequent development of synthetic energetic materials, such as RDX and HMX, pushed performance capabilities to new heights, dramatically altering military and industrial applications, creating the need for newer safety protocols as well. This ongoing evolution reflects humanity's continuous quest for greater control over energy release.

The sensitivity of energetic materials is a critical factor governing their safety and handling. This refers to their susceptibility to detonation from various stimuli such as friction, shock, or impact. High sensitivity materials require extreme caution and specialized handling techniques to prevent accidental explosions. Conversely, less sensitive materials are safer to handle, making them more suitable for specific applications or even the creation of insensitive munitions, a field dedicated to making safer materials for military applications. The balance between high energy output and low sensitivity is a constant challenge for researchers.

The chemical structures of energetic materials play a critical role in determining their properties. Many high-performance energetic materials contain nitrogen-rich molecules. These nitrogen atoms, when released during decomposition, form stable gaseous products such as nitrogen and water, contributing significantly to the energy release. The presence of specific functional groups, such as nitro and nitramine groups, also significantly influences their sensitivity, stability, and overall performance characteristics. Careful design and selection of these structural elements are crucial for achieving desired properties.

The stability of an energetic material is also of paramount importance. It refers to its ability to resist decomposition or unwanted reactions over time and under various environmental conditions. Factors like temperature, humidity, and exposure to UV light can affect the stability of energetic materials. A lack of stability could lead to unpredictable behavior and potential hazards, so ensuring long-term stability is crucial for reliable and safe storage and applications. Indeed, stability studies are an important part of their characterization.

The synthesis of energetic materials involves intricate chemical processes. Many require careful control of reaction conditions such as temperature, pressure, and the addition of specific reagents to ensure purity and high yield. The development of new synthetic routes is a continuous area of research, seeking to improve efficiency, reduce costs, and develop safer and more environmentally friendly methods. Modern synthetic techniques and the use of catalysts are improving efficiency in this field.

The environmental impact of energetic materials is a growing concern. The production, handling, and disposal of many energetic materials can generate toxic byproducts and pollute the environment. Researchers are actively exploring ways to mitigate these impacts by developing "greener" energetic materials with reduced toxicity and lower environmental footprints. This involves finding less hazardous chemicals, improving waste management, and creating sustainable manufacturing practices to reduce their impact on the environment. This represents a shift in the field toward more sustainable practices.

In conclusion, the field of energetic materials is multifaceted, involving chemistry, physics, and engineering disciplines. Understanding their classification, historical development, sensitivity, chemical structures, synthesis, stability, and environmental impact is fundamental for working safely and efficiently with these powerful compounds. Continuous research aims to improve the safety, performance, and environmental compatibility of these materials, shaping their applications across multiple industries.

1.1 Energetic Material Classification

The classification of energetic materials isn't straightforward; it's a multifaceted process requiring consideration of various properties and intended applications. One common approach categorizes them based on their sensitivity to initiation: primary explosives, requiring only a small amount of energy to detonate, like lead azide or mercury fulminate, are often used in detonators. Secondary explosives, in contrast, demand a stronger initiating stimulus; examples include TNT, RDX, and HMX, employed in larger-scale applications such as military ordnance and mining. Their sensitivity differences stem from their chemical structures and inherent instability.

Beyond sensitivity, application plays a critical role. High explosives, mentioned above, generate a rapid and powerful detonation for destructive purposes. Propellants, however, burn more gradually, providing sustained thrust for propulsion systems in rockets, missiles, and firearms. The difference lies in the rate of energy release: a controlled burn for propellants versus an extremely rapid detonation for high explosives. This distinction is crucial for their safe and effective use. Pyrotechnics represent another category, producing light, heat, smoke, or sound through carefully controlled chemical reactions - fireworks being the most familiar example.

Furthermore, we can classify energetic materials by their chemical composition. Nitramines, such as RDX and HMX, are widely used due to their high energy density and performance characteristics. Nitroaromatics, like TNT, also find significant applications, particularly in military explosives. Other classes encompass nitrates, perchlorates, and even less common energetic compounds containing elements like boron, aluminum, or even exotic energetic salts. Each class offers a unique set of performance traits, sensitivities, and production challenges. The selection process necessitates careful consideration of factors including energy output, stability, sensitivity, cost, and environmental impact.

The physical form is also a classifying factor. Solid energetic materials are prevalent in military applications and many commercial pyrotechnics. However, liquid explosives, such as nitroglycerin in its unstabilized form, or more modern liquid formulations used in specific rocket propellants, are also essential, offering unique advantages like ease of handling or increased density. The choice of solid versus liquid significantly impacts the design and handling protocols, requiring specialized storage, transportation, and manufacturing considerations. Gaseous energetic materials, while less common, exist and find niches in specific propulsion systems. Their behavior under pressure, their storage requirements, and potential safety issues are unique to this class.

Ultimately, energetic material classification necessitates a holistic approach considering many interwoven factors. It's not simply about a single property but a complex interplay between sensitivity, application, chemical composition, and physical form. Understanding these classifications is essential for responsible development, safe handling, and efficient application across diverse industries, including military, aerospace, and even mining. Failure to account for these factors can lead to disastrous consequences. Choosing the right energetic material for any given application demands a thorough knowledge of its properties and limitations and carefully considering its compatibility with its surroundings. Each material demands specific handling precautions based on its sensitivity and properties, as failure to do so can result in dangerous outcomes. The entire field rests upon understanding this complexity.

1.2 Historical Overview

The earliest recorded use of energetic materials traces back millennia, to ancient China's invention of gunpowder. This serendipitous discovery, initially used for medicinal purposes, revolutionized warfare and pyrotechnics. The initial mixtures were rudimentary, far from the precisely formulated compounds we understand today. However, this early experimentation laid the groundwork for centuries of development and refinement, a journey marked by both accidental breakthroughs and deliberate scientific investigation. The progression wasn't linear; periods of stagnation followed by rapid advancements shaped the timeline. Early alchemists, while not always driven by the same goals as modern chemists, contributed vital knowledge of chemical reactions and interactions, albeit sometimes shrouded in mysticism.

The European Renaissance witnessed a renewed interest in gunpowder technology, and its military applications fuelled further development. Increased understanding of chemistry and physics allowed for more controlled production and improved formulations. The shift from simply mixing ingredients to carefully controlling stoichiometry marked a crucial step in advancing the potency and reliability of these materials. This period saw the emergence of skilled artisans and early scientists who began to systematically explore the chemical properties of these compounds. The development of more efficient manufacturing processes played a crucial role in scaling up production to meet the growing demands of the increasingly militarized world. This period also saw the beginning of what could be considered the field of energetic materials engineering, with a focus on the optimization of weapon systems and the development of novel devices. The 19th and 20th centuries brought about a massive leap forward. The discovery of new elements and the advancement of organic chemistry led to the synthesis of entirely new, more powerful explosives. Nitrocellulose, nitroglycerin, and later, TNT, revolutionized warfare, influencing strategies and military technologies in profound ways. The development of these compounds wasn't solely about increased destructive power; it also necessitated significant advancements in handling and safety protocols. The inherent dangers of these materials became glaringly apparent, driving research into safer manufacturing methods and storage techniques. This period marked a shift towards a more scientific and systematic approach, moving beyond the empirical methods of the past.

World War I and II significantly accelerated the field's progress, driven by immense military demands. Research and development focused on creating more powerful, stable, and reliable explosives, as well as propellants for both firearms and rockets. This spurred intensive research into the fundamental chemistry of energetic materials and their behavior under various conditions. The war efforts fostered collaborations between chemists, physicists, and engineers, resulting in interdisciplinary advancements. This period saw the synthesis and widespread adoption of RDX and HMX, exceptionally powerful explosives that are still in use today.

Post-war research focused on refining existing formulations, improving safety, and exploring new applications. The development of insensitive munitions, aimed at reducing the risk of accidental detonation, exemplifies this shift towards a more responsible and safety-conscious approach. The field also expanded beyond military applications, encompassing civilian uses such as mining, demolition, and pyrotechnics. The growth in understanding of the chemical and physical properties of energetic materials continued apace, with new tools and techniques from other scientific disciplines, such as spectroscopy and computational chemistry, expanding the scope of research. This interdisciplinary effort has led to the ongoing evolution of safer, more environmentally friendly compounds.

The late 20th and early 21st centuries have seen a significant increase in the application of computational chemistry and advanced material science techniques. This allows for the theoretical prediction of the properties of new energetic materials before engaging in lengthy and potentially dangerous experimental work. This capability not only accelerates the development process, but also reduces the environmental impact by minimizing the amount of experimental work needed, and thus lessening the risk of hazardous waste production. These techniques, combined with a deeper understanding of reaction mechanisms, provide insights into the design of high-performance, environmentally friendly, and safer energetic materials. The future of energetic materials research promises exciting advancements in safety, performance, and sustainability.

2 Thermochemistry of Energetic Materials

Understanding the thermochemistry of energetic materials is paramount to designing, handling, and utilizing these powerful compounds safely and effectively. The energy released during their decomposition, a defining characteristic, is fundamentally governed by thermodynamic principles. This release isn't simply a matter of explosive power; it's a complex interplay of heat, energy, and the speed at which the reactions occur.

Heat of formation, a key thermodynamic property, represents the enthalpy change associated with the formation of one mole of a substance from its constituent elements in their standard states. For energetic materials, this value is often negative, indicating an exothermic reaction - heat is released upon formation. A more negative heat of formation suggests a higher energy density, implying a greater potential for energy release during decomposition. Precise measurements of heats of formation, usually obtained through sophisticated calorimetry techniques, are essential for predicting the performance of energetic compounds. These values directly influence the energy output and therefore the overall efficacy of the explosive or propellant.

Enthalpy and entropy are inseparable in dictating the spontaneity and equilibrium of chemical reactions. Enthalpy (H), representing the total heat content of a system, is a crucial factor in determining the energy released during an explosion. A highly negative change in enthalpy (H) signifies a highly exothermic reaction, crucial for the desired energetic output. Entropy (S), reflecting the randomness or disorder of a system, also plays a vital role. While an increase in entropy usually favors a reaction, the effect is less direct than that of enthalpy in high-energy material reactions. The Gibbs free energy (G), defined as G = H \item TS (where T is the temperature), ultimately determines the spontaneity of a reaction, providing valuable insights into the stability and reactivity of energetic materials at various temperatures.

Reaction kinetics, the study of reaction rates and mechanisms, provide a crucial understanding of how quickly and by what pathways energetic materials decompose. These processes are often complex, involving multiple steps and intermediate species. The activation energy (Ea), the minimum energy required for a reaction to proceed, is a key parameter. A lower activation energy implies a faster reaction rate, leading to a more rapid energy release. Understanding the kinetics allows prediction of the burn rate of propellants or the detonation velocity of explosives. Factors influencing reaction rates include temperature, pressure, and the presence of catalysts or inhibitors, all of which need meticulous consideration in the design and safe handling of these materials. Advanced techniques like

Differential Scanning Calorimetry (DSC) and Thermogravimetric Analysis (TGA) provide essential data to characterize and model the kinetics of energetic material decomposition.

Predicting the performance of an energetic material requires integrating thermochemical data with kinetic information. Modeling software employing complex algorithms simulates the decomposition process, combining thermodynamic properties with the kinetic rates to estimate the energy released, pressure generated, and other critical parameters. Accurate modeling relies on precise experimental data and sophisticated theoretical frameworks to accurately forecast real-world performance and ensure consistency with test results. Combining experimental and computational methods helps optimize the energetic material's properties for specific applications.

Furthermore, the sensitivity and stability of energetic materials are closely tied to their thermochemical properties. Materials with low activation energies and high energy densities can be more sensitive to accidental initiation, requiring careful consideration during manufacture, handling, storage, and transportation. Understanding the interplay between thermodynamic parameters and sensitivity provides crucial information for the design of safer, more stable energetic materials. Ultimately, a deep understanding of the thermochemistry of energetic materials is not just crucial for optimizing performance, but also paramount for ensuring their safe and responsible use. This knowledge aids in risk mitigation and guides the development of safer formulations for military and industrial applications.

2.1 Heat of Formation

Precise determination of heat of formation is crucial for understanding and predicting the energetic behavior of materials. This value, representing the enthalpy change associated with the formation of one mole of a compound from its constituent elements in their standard states, directly impacts the material's overall energy density. A higher heat of formation often indicates a greater potential for energy release upon decomposition, a critical factor in assessing the performance of explosives and propellants. Therefore, accurate experimental methods and sophisticated computational techniques are vital in determining these values.

Numerous experimental techniques exist for measuring heats of formation, each with its strengths and limitations. Calorimetry, for example, involves measuring the heat absorbed or released during a chemical reaction. Different types of calorimetry, such as combustion calorimetry or solution calorimetry, are employed depending on the specific energetic material and its properties. Bomb calorimetry, a common technique for determining the heat of combustion, often serves as a crucial step in calculating the heat of formation via Hess's Law, which involves using known heats of combustion and formation of related compounds to derive the desired value. However, experimental challenges remain, including the potential for incomplete combustion or side reactions, which can lead to inaccuracies in the final results.

Computational chemistry offers a powerful alternative to purely experimental approaches, especially when dealing with unstable or hazardous compounds. Methods such as density functional theory (DFT) allow for the prediction of heats of formation through theoretical calculations. While computationally intensive, these methods provide valuable insights into the electronic structure and bonding within the energetic material, offering a deeper understanding of the factors that influence its energetic properties. Comparison between experimentally determined and computationally predicted heats of formation not only validates computational methods but also illuminates the underlying chemistry governing energetic behavior. However, the accuracy of computational predictions relies heavily on the choice of theoretical method and basis set employed. Careful consideration and validation against experimental data are necessary to ensure reliable results.

The heat of formation is not solely a fundamental thermodynamic property but has considerable practical implications. It is a key parameter in the design and optimization of new energetic materials. By understanding the relationship between molecular structure and heat of formation, researchers can tailor the design of molecules to achieve desired energetic performance. This includes factors such as the energy density, sensitivity, and stability of the material. For example, incorporating specific functional groups or altering the molecular architecture can significantly impact the heat of formation, thereby influencing the overall performance characteristics of the energetic material. This is crucial for balancing the required energy output with the necessary safety and stability requirements.

Furthermore, the heat of formation plays a crucial role in assessing the relative stability of different energetic materials. A more negative heat of formation generally indicates a greater thermodynamic stability. While kinetic factors, such as activation energies, can significantly influence the rate of decomposition, the heat of formation provides valuable insights into the overall energy content and potential for energetic release. This information is essential for predicting the potential hazards associated with handling and storing various energetic materials and for implementing appropriate safety measures. Understanding these fundamental thermodynamic properties, such as the heat of formation, is not just an academic pursuit; it is vital for ensuring safe handling, reliable performance and responsible development of energetic materials. The synergistic approach of combining experimental measurements with sophisticated computational methods remains a vital step in characterizing the complex energetic behavior of these critical materials. Accurate heats of formation serve as a cornerstone in the design, development, and application of these powerful substances. A thorough and comprehensive investigation of heats of formation, both through experimental and computational techniques, is essential for responsible advancement in the field of energetic materials. These insights underpin the creation of safer, more efficient, and more environmentally friendly energetic materials. Continued research and refinement of these methodologies will continue to advance our understanding and ability to safely utilize the transformative power of energetic materials.

2.2 Enthalpy and Entropy

Understanding enthalpy and entropy is crucial for predicting the energetic performance of materials. Enthalpy (H), a measure of a system's total heat content, directly relates to the energy released or absorbed during a reaction. A negative change in enthalpy (H < 0) indicates an exothermic reaction—one releasing heat, which is desirable in energetic materials. Conversely, a positive H signifies an endothermic reaction, absorbing heat. The magnitude of H reflects the amount of energy released or absorbed, directly impacting the power of the explosive. Precise calculations require careful consideration of standard enthalpy of formation values for all reactants and products, often found in thermochemical databases.

Entropy (S), on the other hand, quantifies the randomness or disorder within a system. A higher entropy indicates greater disorder. In energetic materials, the change in entropy (S) during a reaction influences the spontaneity of the process. For a reaction to be spontaneous at constant temperature and pressure, the Gibbs free energy (G) must be negative. Gibbs free energy is defined by the equation G = H \item TS, where T is the absolute temperature. Thus, even if a reaction is exothermic (negative H), a large negative S might render the reaction non-spontaneous at lower temperatures, highlighting the importance of considering both enthalpy and entropy. Accurate entropy calculations require detailed knowledge of the system's structure, and sophisticated computational methods are often employed.

The relationship between enthalpy and entropy is often counterintuitive in the context of energetic materials. For instance, while high explosives aim for maximum energy release (large negative H), the resulting products might possess a lower entropy than the reactants. This is because a complex molecule breaks down into smaller, more stable fragments, reducing overall disorder. However, this decrease in entropy often isn't significant enough to counteract the large exothermic nature of the reaction. The reaction proceeds spontaneously due to the dominant influence of the highly negative enthalpy change.

Consequently, optimizing energetic materials involves balancing enthalpy and entropy. A high enthalpy ensures substantial energy release, while a favorable entropy change enhances the reaction's spontaneity. In practice, this often translates to searching for compounds with high heat of formation (highly positive Hf) and a manageable entropy decrease during the decomposition reaction. The balance, however, will vary depending on the specific application. For instance, propellants might prioritize controlled energy release over maximum power, requiring a different enthalpy-entropy balance than that of high explosives.

Experimental determination of enthalpy and entropy often involves calorimetric techniques, such as differential scanning calorimetry (DSC) and bomb calorimetry. DSC measures the heat flow associated with phase transitions and chemical reactions, providing insights into enthalpy changes. Bomb calorimetry, a more sophisticated technique, determines the heat of combustion, providing another route to calculate enthalpy changes. Entropy values are often obtained indirectly through thermodynamic relationships or directly using statistical mechanical methods employing molecular modelling and simulations. These experimental and computational methods provide essential data for accurate predictions and design of new energetic materials.

Furthermore, computational chemistry methods have become increasingly important tools in predicting enthalpy and entropy. These methods, including density functional theory (DFT) calculations, can estimate the energy and structure of molecules, providing valuable data for predicting reaction energetics. These computations help researchers screen numerous potential candidates before undertaking expensive and time-consuming experimental synthesis and characterization. This computational approach significantly accelerates the discovery and development of novel high-performance energetic materials.

It is crucial to remember that the accuracy of enthalpy and entropy predictions hinges on the accuracy of the employed methodologies and the availability of relevant experimental data. Therefore, a collaborative approach combining experimental measurements and advanced computational modeling is often necessary for achieving accurate and reliable results. Understanding these limitations is crucial in interpreting the results and making informed decisions during the design and development process. The interplay between theory and experiment is paramount for success in this field.

In summary, mastering the concepts of enthalpy and entropy is not merely academic; it is pivotal in understanding and designing energetic materials. A deep understanding of these thermodynamic properties, coupled with the capabilities of modern computational tools and experimental techniques, empowers researchers and engineers to develop safer, more powerful, and more environmentally responsible energetic materials for various applications. The ability to predict and optimize these properties will continue to be critical in the advancement of this field for years to come. The pursuit of this knowledge continues to shape the landscape of energetic material research, pushing the boundaries of safety and performance.

2.3 Reaction Kinetics

Understanding reaction kinetics is crucial for predicting the performance and safety of energetic materials. The decomposition of these materials is often complex, involving multiple parallel and sequential reactions. Kinetic parameters, such as activation energies and reaction orders, govern the rate at which these reactions proceed, directly influencing the energy release rate, explosion velocity, and overall effectiveness of the compound. Precise determination of these parameters is vital, impacting the design of safer and more efficient energetic materials.

The study of reaction kinetics relies heavily on experimental techniques. Differential scanning calorimetry (DSC) and thermogravimetric analysis (TGA) are frequently employed to determine the thermal decomposition behavior of energetic materials. These methods provide valuable data on the reaction rates at various temperatures. By analyzing the resulting data using appropriate kinetic models, we can extract key parameters like the activation energy (Ea), a measure of the energy barrier that must be overcome for the reaction to proceed. A lower activation energy implies a faster reaction rate, increasing the potential hazard in the case of energetic materials.

Furthermore, the reaction order (n) describes how the reaction rate depends on the concentration of the reactants. This parameter is essential for accurate modelling of the decomposition process. For instance, a first-order reaction (n=1) indicates that the rate is directly proportional to the concentration, while a second-order reaction (n=2) shows a dependence on the square of the concentration. The identification of the correct reaction order is critical for accurate prediction of the reaction rate under varying conditions. Methods for determining reaction order and activation energy often involve non-linear least-squares fitting of experimental data to various kinetic models, such as the Arrhenius equation, which relates the reaction rate constant to temperature and activation energy.

The influence of environmental factors on reaction kinetics cannot be overlooked. Factors like pressure, humidity, and the presence of catalysts or inhibitors can significantly alter the reaction rates and pathways. For example, elevated pressure can accelerate decomposition processes, while certain additives can act as catalysts, lowering the activation energy and increasing the reaction rate. Conversely, inhibitors can slow down decomposition, enhancing safety. A thorough understanding of these factors is crucial for predicting the behavior of energetic materials under various operational conditions and for designing formulations with improved performance and safety characteristics.

Moreover, detailed kinetic studies also offer insight into the reaction mechanisms. This involves identifying the intermediate species and determining the sequence of elementary reactions that constitute the overall decomposition process. Employing techniques such as mass spectrometry and gas chromatography can help analyze the gaseous products generated during decomposition, revealing valuable information about the reaction pathways. This mechanistic understanding allows for more refined kinetic modeling and ultimately, improved control over the performance and safety of energetic materials.

Finally, accurate kinetic modeling is essential for designing safer and more efficient energetic materials. By understanding the reaction kinetics, we can predict the behavior of materials under various conditions, enabling the optimization of their performance characteristics and reducing the risk of accidental detonation or other hazards. This knowledge allows for the development of materials with tailored energy release profiles, enhanced stability, and reduced sensitivity, directly contributing to the development of more advanced and reliable energetic systems. This detailed kinetic analysis allows for creating energetic materials with improved performance and enhanced safety profiles. In essence, understanding reaction kinetics is not just a theoretical exercise; it is an indispensable tool for the design and implementation of safe and efficient energetic materials for various applications. The integration of experimental data analysis and sophisticated kinetic modelling creates a powerful framework for predictive capabilities crucial for innovation in this field.

3 Nitramines and Nitroaromatics

RDX and HMX, cyclic nitramines, are cornerstones of modern explosives. Their synthesis, often involving nitration of hexamine, presents a fascinating blend of organic chemistry and industrial process engineering. Careful control of reaction conditions, including temperature and acid concentration, is crucial for maximizing yield and purity, minimizing unwanted byproducts like nitroglycerin. Understanding the intricate reaction mechanisms is key to optimizing these high-energy density materials. The precise control of crystallization processes is also crucial for achieving the desired crystal morphology, impacting sensitivity and performance.

The iconic TNT, or 2,4,6-trinitrotoluene, possesses a unique combination of properties that have solidified its role as a benchmark explosive. Its relatively low sensitivity to shock and friction, combined with its high melting point, facilitates safe handling and casting, making it suitable for a variety of applications, from military munitions to demolition work. However, TNT's toxicity and environmental persistence are significant drawbacks, motivating research into less harmful alternatives. The complex interplay between

TNT's molecular structure and its properties, particularly its detonation characteristics, makes it a subject of ongoing scientific investigation. Detailed analyses of its crystal structure reveal crucial information regarding its stability and sensitivity.

Nitroguanidine, a less powerful but valuable energetic material, showcases versatility beyond its explosive properties. Its primary application lies in propellant formulations, where it contributes to burning rate control and overall performance. Its relatively low sensitivity makes it a safe component in many propellant blends. Furthermore, the unique characteristics of nitroguanidine's decomposition chemistry allow for precise tailoring of the burning characteristics in solid rocket propellants, influencing factors like thrust profile and overall efficiency. Its use demonstrates the ability to harness the energy stored in seemingly less powerful materials for specific applications.

The sensitivity and stability of nitramines and nitroaromatics are paramount considerations, dictating handling procedures and storage requirements. Sensitivity to friction, impact, and electrostatic discharge varies significantly depending on the specific compound, crystal morphology, and presence of impurities. Stability, on the other hand, relates to the material's propensity to undergo unwanted chemical changes over time, such as decomposition or isomerization, influencing its long-term performance and safety. This emphasizes the need for rigorous quality control during synthesis and storage. Understanding the factors that influence sensitivity and stability is essential for designing safe and reliable energetic materials for a wide range of applications.

The inherent toxicity and environmental impact of these energetic materials are increasingly important concerns. Many nitramines and nitroaromatics are known to be toxic to humans and other organisms, raising concerns about occupational exposure and environmental contamination. TNT, for instance, is a known carcinogen. Disposal methods require careful consideration, balancing the necessity of safe and efficient disposal with cost considerations. The environmental fate and potential for bioaccumulation in soil and water need careful study, motivating the exploration of bioremediation strategies and environmentally friendlier alternatives. Sustainable synthesis and degradation methods are essential for reducing the environmental burden.

3.1 RDX and HMX Synthesis

The synthesis of RDX (cyclotrimethylenetrinitramine) and HMX (cyclotetramethylenetetranitramine) represents a cornerstone of energetic material production. These powerful nitramines find extensive use in military applications, demanding precise control over their synthesis to ensure both high yield and purity. The methods employed are typically based on nitration reactions, leveraging the powerful oxidizing properties of nitric acid. However, the specific conditions and procedures vary significantly.

A common pathway to RDX involves the nitration of hexamine. This process, often conducted in a mixture of concentrated nitric acid and acetic anhydride, carefully balances reaction temperature and time to maximize

RDX formation and minimize undesired byproducts. Efficient heat removal is critical to avoid runaway reactions, while meticulous control of the reaction parameters is essential for achieving high yields of pure RDX. Careful attention must be paid to the safety protocols, given the hazardous nature of the reactants and the potential for exothermic decomposition. Variations in the nitration process, including the use of different solvents or the addition of catalysts, can impact the product's crystalline structure and particle size distribution. These factors significantly influence the performance characteristics of the resulting RDX-based explosives.

In contrast to RDX's synthesis from hexamine, the production of HMX often employs a more complex route. While direct nitration of hexamine can yield small quantities of HMX, optimized synthesis usually involves controlled nitration of hexamine under specific conditions or even the use of other precursor molecules, frequently involving a multi-step approach. The influence of reaction parameters like temperature, acid concentration, and the presence of specific additives are even more critical in the case of HMX synthesis. Moreover, the purification of HMX often presents a more challenging task due to its inherent similarity to other nitramine byproducts. The isolation and crystallization processes therefore require careful optimization to ensure a product of suitable purity and quality for intended military applications.

The selection of a specific synthetic route depends heavily on factors such as desired yield, purity requirements, and the availability of resources. Each method presents its own challenges and advantages. For instance, while the hexamine route to RDX is relatively straightforward, the yield can be sensitive to reaction conditions. Conversely, the more intricate pathways for HMX synthesis can offer potentially higher purity but at the expense of increased complexity and greater potential for safety hazards. Industrial-scale production necessitates sophisticated process control and safety measures, adhering strictly to rigorous safety guidelines and environmental regulations.

Beyond the core synthesis techniques, significant research continues to explore alternative approaches, aiming for improved efficiency, reduced environmental impact, and enhanced safety. This research incorporates various approaches, including the investigation of new catalysts, the exploration of greener solvents, and the refinement of existing processes to minimize waste and energy consumption. Continuous optimization of existing methods remains crucial for the manufacturing of high-quality, high-performance RDX and HMX, addressing the ongoing demands for reliable and safe energetic materials in various military and industrial applications. The development of novel synthetic strategies remains a significant area of ongoing investigation, driven by the pursuit of higher yields, better control of product characteristics, and improved sustainability. The efficient and safe production of RDX and HMX remains paramount to ensure both performance and safety in their military applications, a goal actively pursued through continuous research and process improvement. This ongoing work underpins the development and improvement of energetic materials and technologies. The advancements in this field continuously enhance the efficiency, safety, and environmental responsibility of manufacturing these critical compounds. Furthermore, the constant drive for optimization underscores the importance of balancing production efficiency with safety and minimizing environmental impact. These factors are critical to the continued development and responsible use of RDX and HMX within the broader context of energetic materials.

3.2 TNT Properties

2,4,6-Trinitrotoluene, more commonly known as TNT, is a yellowish crystalline solid, a nitroaromatic compound that holds a significant place in the history and ongoing applications of energetic materials. Its relatively low sensitivity to shock and friction, coupled with its ease of handling and melting point conducive to casting, makes it a versatile and historically prevalent choice for military and industrial applications. This doesn't, however, negate the importance of understanding its properties thoroughly before any application.

The chemical formula of TNT is CHNO, reflecting its molecular structure—a toluene molecule (methylbenzene) substituted with three nitro groups (-NO). These nitro groups are the key to TNT's energetic properties, providing the oxygen needed for a self-sustaining reaction upon detonation. The three nitro groups are located at positions 2, 4, and 6 on the benzene ring, maximizing the molecule's symmetry and influencing its crystal packing, thereby indirectly affecting its sensitivity and stability. Understanding this molecular structure is paramount to grasping the compound's overall behavior.

A crucial aspect of TNT's properties centers around its detonation characteristics. Unlike many other high explosives, TNT's detonation velocity is relatively modest, typically around 6900 meters per second. This, combined with its density, around 1.65 g/cm^3, contributes to its overall performance in explosive applications. The detonation pressure generated is significant, but not as extreme as some other high explosives, leading to its classification as a secondary explosive, requiring a primary explosive for initiation. It's important to remember these parameters are influenced by factors such as temperature and purity.

TNT's stability is another critical property. While less sensitive than some primary explosives, it's still hazardous to handle carelessly. It is relatively insensitive to impact and friction, but prolonged exposure to high temperatures or significant increases in pressure can lead to decomposition or even accidental detonation. Careful storage in controlled environments is therefore crucial, preventing premature decomposition and ensuring safety. The stability is crucial not only for safety but also for the predictability and consistency of performance in various applications.

The melting point of TNT, around 80.8 °C, is advantageous for manufacturing and casting. This allows for the creation of shaped charges and the filling of shells, bombs, and other munitions with molten TNT. This characteristic significantly enhances its practical use in military and commercial blasting operations, simplifying the production process and ensuring controlled deployment of the explosive material. Moreover, this property contributes significantly to the ease of handling.

Beyond its energetic properties, the toxicity and environmental impact of TNT are significant concerns. TNT is known to be toxic to both humans and the environment. Long-term exposure can cause anemia, liver damage, and other health problems. Its degradation products can also contaminate soil and water sources, posing environmental hazards requiring specialized remediation techniques. Understanding these aspects is crucial for implementing safe handling procedures and environmentally responsible disposal methods.

The practical applications of TNT continue to evolve, primarily in controlled demolition and mining, although its military use remains relevant. This reflects the complex balance between its powerful explosive properties and its relatively manageable sensitivity, alongside the ongoing efforts to mitigate its environmental and health consequences. It is imperative to maintain awareness of these factors throughout all handling and usage.

TNT remains a subject of ongoing research. Studies focus on improving its safety and stability, reducing its toxicity, and exploring alternative materials to minimize its environmental footprint. This research emphasizes the significance of responsible development and application of energetic materials, a commitment to safety and sustainability.

In conclusion, 2,4,6-trinitrotoluene, despite its hazardous nature, retains an important position in the field of energetic materials due to its versatile properties. Understanding its molecular structure, detonation characteristics, stability, melting point, and toxicity is essential for responsible application and safe handling. Ongoing research into safer alternatives and mitigation strategies emphasizes the continuous need for responsible development and use of such powerful compounds. These properties highlight the complex nature of energetic materials chemistry and its significant impact across diverse fields. The safe and responsible use of TNT depends on a comprehensive understanding of these various characteristics and their interrelation.

3.3 Nitroguanidine Applications

Nitroguanidine, a relatively insensitive energetic material, finds diverse applications despite its lower explosive power compared to compounds like RDX or TNT. Its unique properties, particularly its low sensitivity to impact and friction, make it a valuable component in various formulations. This inherent safety characteristic is crucial in minimizing risks during manufacturing, handling, and storage. The relatively low cost of nitroguanidine production further enhances its attractiveness for large-scale applications. Unlike many other energetic materials that primarily serve as high explosives, nitroguanidine's role often focuses on modifying the performance characteristics of existing formulations. It acts as a powerful depressant, lowering the burning rate of propellant mixtures. This control over burn rate is critical in tailoring propellants for specific applications, such as adjusting the thrust profile of rockets or ensuring consistent performance in firearms. The ability to modulate burn rate without significantly impacting the overall energy output allows for greater precision in propellant design.

The incorporation of nitroguanidine in propellant compositions helps to improve their overall safety profile. Its stabilizing influence minimizes the risk of accidental detonation, especially during storage or transportation. This is particularly important in military applications where safety is paramount. For example, in solid rocket propellants, nitroguanidine contributes to a more predictable and controlled combustion process, reducing the likelihood of uncontrolled explosions or deflagrations. This enhanced safety feature significantly reduces the potential for accidents and minimizes the risk to personnel.

Beyond propellants, nitroguanidine's applications extend to other specialized areas within the field of energetic materials. It often serves as a crucial ingredient in specialized pyrotechnic compositions where its low sensitivity is highly beneficial. In these formulations, nitroguanidine contributes to controlled burning and helps create predictable visual effects, such as colored flames or smoke generation. This careful control is vital for achieving specific effects in pyrotechnic displays and signaling devices. Its impact and friction insensitivity make it especially suitable for applications requiring enhanced safety in crowded areas.

Furthermore, the incorporation of nitroguanidine into explosive compositions can help to moderate the sensitivity of other, more reactive materials, thereby producing a safer and more stable mixture. The blend of nitroguanidine's insensitivity and energetic capabilities offers a unique balance for specific applications where a lower intensity, more controllable explosive effect is necessary. This approach provides a path to creating specialized mixtures tailored to exact requirements, such as those found in demolition work, mining, or certain industrial processes.

The unique blend of safety, relatively low cost, and controllable energetic properties make nitroguanidine a valuable and versatile component in a wide range of energetic material applications. Its inclusion often improves overall safety, offers precise control over burn rates, and enables the creation of customized explosive and pyrotechnic compositions. This makes it an integral material in both military and civilian applications, underscoring its continued importance in the field. Ongoing research continues to explore new and innovative applications for nitroguanidine, extending its usefulness and expanding its importance within the world of energetic materials chemistry. Research into its incorporation into greener formulations also aims to reduce the environmental impact associated with energetic material use, a critical concern for future generations. This ongoing development of nitroguanidine applications reflects its enduring role within the broader field of energetic materials.

3.4 Sensitivity and Stability

The sensitivity and stability of energetic materials are paramount concerns, dictating their safe handling, storage, and practical application. A compound's sensitivity reflects its susceptibility to initiation by external stimuli such as friction, impact, or electrostatic discharge. Conversely, stability refers to a material's resistance to decomposition or unwanted reactions over time, under various environmental conditions. These two properties are intrinsically linked; a highly sensitive material often possesses limited stability, and vice versa. Understanding this complex interplay is crucial for designing and using energetic materials responsibly.

Nitroaromatics, like TNT, are generally less sensitive than nitramines such as RDX and HMX. However, their stability can vary widely depending on purity, storage conditions, and the presence of impurities. Factors like crystal morphology, particle size, and the presence of defects within the crystal lattice all impact a material's sensitivity. For instance, smaller particle sizes often lead to increased sensitivity due to the larger surface area available for interaction with external stimuli. Similarly, the presence of even trace amounts of impurities can significantly alter a compound's sensitivity and stability, potentially leading to unexpected and hazardous behavior.

Designing insensitive munitions, therefore, requires careful consideration of both these aspects. Techniques include reducing crystal size, controlling the morphology of the energetic crystals, and employing desensitizing additives. These additives may act as physical barriers, reducing friction and impact sensitivity, or as chemical inhibitors, slowing down decomposition reactions. The selection of appropriate additives must balance the desire for reduced sensitivity with any potential negative impacts on the material's performance characteristics. It's a delicate balance. Careful investigation into the chemical interactions between the energetic material and the additive is paramount to avoid unexpected consequences. The goal is improved safety without compromising effectiveness.

Moreover, the chemical environment significantly influences the stability of energetic materials. Exposure to elevated temperatures, humidity, or UV radiation can accelerate decomposition, potentially leading to the formation of unstable by-products. This decomposition can often generate heat which can subsequently trigger further decomposition, potentially leading to runaway reactions. Therefore, rigorous testing and quality control are essential throughout the production and storage processes to ensure material integrity and minimize the risk of accidents. Environmental conditions must be controlled precisely to maintain optimal stability and prevent premature degradation. This control goes beyond simple temperature and humidity regulation, encompassing all aspects of the environment that could impact the material's properties.

Furthermore, understanding the underlying chemical mechanisms responsible for sensitivity and stability is vital for designing improved energetic materials. Detailed kinetic studies can provide insights into the reaction pathways involved in decomposition, enabling the identification of potential vulnerabilities or points of instability. This information can then be used to tailor synthesis methods or modify the molecular structure of the energetic material to enhance its stability and reduce its sensitivity. This refined approach allows for the creation of materials that meet the demands of specific applications while prioritizing safety. We must prioritize this approach; it is a matter of life and safety.

The development of new energetic materials requires a holistic approach, integrating principles of thermodynamics, kinetics, and materials science. This integration allows for the rational design of compounds with predictable sensitivity and stability characteristics. Computational modeling techniques are increasingly utilized to predict the properties of novel compounds before synthesis, reducing the need for extensive experimental testing and promoting efficient use of resources. This predictive capability allows us to investigate numerous possibilities and to optimize designs quickly, a significant leap forward for energetic materials research.

Ultimately, ensuring the sensitivity and stability of energetic materials is an ongoing challenge that requires a multidisciplinary approach. It demands constant collaboration and an ever-increasing understanding of the intricate relationships between molecular structure, thermodynamic properties, and physical behavior. The commitment to safety should be integrated into every stage of the process, from synthesis and formulation to handling and disposal. Ongoing research and development, coupled with stringent safety protocols, are crucial in mitigating risks and maximizing the safe and effective utilization of these powerful materials. The future of energetic materials research lies not merely in increasing power, but in achieving this power with enhanced safety.

3.5 Toxicity and Environmental Impact

The inherent toxicity of many energetic materials poses a significant challenge. Compounds like TNT, RDX, and HMX, while possessing desirable energetic properties, are known to be harmful to human health and the environment. TNT, for instance, is a potent mutagen and carcinogen, readily absorbed through skin contact, inhalation, or ingestion, leading to a range of severe health problems, including liver damage, anemia, and leukemia. Exposure to even low levels over extended periods can have devastating consequences, highlighting the need for stringent safety protocols in manufacturing, handling, and disposal. This underscores the critical importance of developing safer alternatives. The insidious nature of these toxins makes remediation efforts complex and costly, demanding innovative solutions.

RDX and HMX, while less acutely toxic than TNT, still present considerable environmental concerns. Their persistence in soil and water bodies, coupled with their potential to leach into groundwater, poses a long-term threat to ecological systems. These compounds can disrupt soil microbial communities, affecting nutrient cycling and plant growth. Their accumulation in aquatic environments can cause harm to aquatic organisms, particularly fish and invertebrates, through bioaccumulation and biomagnification. The impact extends beyond individual organisms, affecting entire ecosystems. Detailed studies using advanced analytical techniques such as chromatography-mass spectrometry (GC-MS and LC-MS) are essential to monitor their distribution and assess their long-term effects. This data is crucial for informed decision-making in environmental risk assessment and mitigation.

The environmental impact extends beyond the compounds themselves. The manufacturing processes often involve the use of hazardous solvents and chemicals, contributing to air and water pollution. Wastewater from production facilities may contain residual energetic materials and their byproducts, requiring specialized treatment to prevent environmental contamination. The explosion of energetic materials, whether accidental or intentional, can release toxic fumes and particulate matter into the atmosphere, causing short-term respiratory problems and potentially long-term health effects. Explosions also cause physical disruption, leaving behind contaminated soil and debris that requires careful remediation. This encompasses various cleanup strategies, from simple excavation and disposal to advanced bioremediation techniques, necessitating thorough site assessments and tailored remediation plans.

Therefore, a multifaceted approach to mitigate the toxicity and environmental impact is necessary. This involves stricter regulations on the manufacturing, storage, and disposal of energetic materials, as well as the development and implementation of cleaner production technologies. Research into less toxic energetic materials is crucial, focusing on compounds with reduced environmental persistence and lower toxicity profiles. Green chemistry principles should guide the design and synthesis of new energetic materials, prioritizing the use of less hazardous solvents and reagents and minimizing waste generation. Furthermore, the development of innovative remediation technologies is vital for addressing existing contamination. These technologies should be both effective and cost-effective, focusing on long-term solutions and sustainable environmental management.

Beyond technological advancements, effective regulatory frameworks and comprehensive risk assessment protocols are crucial. These include stringent environmental regulations, robust monitoring programs, and public awareness campaigns to educate individuals and communities about the risks associated with energetic materials. The success of mitigation strategies relies on a holistic approach that combines scientific innovation, policy measures, and responsible stewardship. International collaboration is also essential, fostering the sharing of knowledge, best practices, and innovative technologies to address this global challenge. This collaborative effort, encompassing governments, industries, and research institutions, is critical for protecting human health and preserving the environment. The ultimate goal is to achieve a balance between national security needs and environmental protection, ensuring the responsible use and sustainable management of energetic materials. This necessitates a continuous cycle of innovation, evaluation, and refinement, ensuring that future advancements in energetic materials chemistry prioritize both effectiveness and environmental sustainability.

4 High Explosives: Synthesis and Properties

The distinction between primary and secondary high explosives hinges primarily on their sensitivity to initiation. Primary explosives, such as lead azide or mercury fulminate, are extremely sensitive to friction, impact, or heat, detonating readily with minimal stimulus. Their use is typically confined to detonators or primers, initiating the detonation of less sensitive secondary explosives. Their chemical structures often feature strained rings or unstable functional groups, contributing to their inherent instability and proneness to rapid decomposition. Careful handling and specialized storage conditions are crucial due to their dangerous reactivity. Understanding their synthesis, involving intricate organic chemistry techniques and often meticulous purification steps, is essential for controlling their performance and stability.

Secondary explosives, conversely, require a significant external stimulus to detonate. Compounds like RDX (cyclotrimethylenetrinitramine), HMX (cyclotetramethylenetetranitramine), and TNT (trinitrotoluene) exemplify this class. Their synthesis is generally more robust and scalable than that of primary explosives, often involving nitration reactions under carefully controlled conditions. These methods, ranging from simple batch processes to continuous flow techniques, are heavily influenced by factors such as temperature, concentration, and the use of specific catalysts. The chemical stability of secondary explosives is significantly greater than that of their primary counterparts, permitting relatively safe handling and storage under appropriate conditions. However, their sensitivity to impact and friction, albeit lower than that of primary explosives, necessitates stringent safety procedures during manufacturing, transportation, and handling.

The inherent energetic properties of these explosives stem from their molecular structure and the presence of specific functional groups. Nitro groups (-NO2), for instance, are prevalent in many high explosives. The presence of multiple nitro groups, alongside other factors like the presence of strained rings or other energetic functional groups, enhances the energy density and consequently the explosive power of the compound. This energy density is often quantified through measurements of the heat of detonation and velocity of detonation, parameters crucial for performance prediction and application optimization. These measurements, conducted in controlled laboratory settings using sophisticated equipment, provide essential data for the design and development of munitions and other applications. This careful measurement and analysis are vital in ensuring safety and predictability in using these compounds.

Moreover, the physical properties of high explosives, such as density, crystal structure, and particle size, significantly impact their performance and handling characteristics. High density increases the energy density and enhances the explosive power. The crystalline structure influences sensitivity and stability, with certain crystal forms exhibiting higher sensitivity than others. Particle size distribution affects the ease of handling and the efficiency of detonation, highlighting the importance of precise control over the physical properties during the manufacturing process. It's the careful manipulation of both chemical synthesis and physical properties that leads to reliable and predictable explosives, fit for their intended purpose, whether it be military applications, demolition, or other niche uses. This precise control and understanding are paramount to avoiding catastrophic accidents.

Finally, the development of new high explosives continues to be an active area of research. Focus on increasing energy density while simultaneously reducing sensitivity is driving innovation, aiming for safer and more efficient energetic materials. Computational methods and advanced synthesis techniques are critical to this process, allowing researchers to predict the properties of novel compounds before undertaking the often dangerous and costly experimental synthesis. This computational prediction of both thermodynamic and kinetic properties is increasingly accurate, leading to significant cost and time savings in research and development. Furthermore, ongoing research targets the development of environmentally friendlier alternatives to conventional explosives, reducing the impact of these materials on the environment and human health. The future of high explosives relies on careful balancing of safety, performance, and environmental considerations.

4.1 Primary Explosives

Primary explosives are characterized by their extreme sensitivity to heat, shock, or friction, initiating detonation with minimal external stimulus. Unlike secondary explosives, which require a significant detonation wave for initiation, primary explosives are incredibly unstable and readily decompose, often violently. This inherent instability makes them crucial for initiating larger, less sensitive explosives in various applications, from detonators in mining operations to the firing pins in firearms. Their sensitivity, however, also necessitates careful handling and storage, demanding rigorous safety protocols.

The chemical structures of primary explosives often feature strained ring systems or electron-withdrawing groups, contributing to their energetic instability. Classic examples include lead azide, a highly sensitive compound often used in detonators due to its reliable initiation properties and relatively easy production process. Another common example is mercury fulminate, historically favored for its effectiveness and relative ease of manufacture, though concerns regarding its toxicity have led to its reduced use in recent times. These compounds' inherent reactivity stems from their molecular architecture, which facilitates rapid decomposition into stable products, releasing significant energy in the process.

Modern advancements have focused on developing primary explosives with improved safety characteristics. While retaining their crucial initiation properties, researchers are continually striving to minimize sensitivity while maintaining sufficient explosive power. This ongoing research involves exploring novel chemical structures and synthesizing compounds that exhibit lower sensitivity to external stimuli while preserving or enhancing their detonation capabilities. These efforts leverage advanced analytical techniques to fine-tune the synthesis and handling procedures, ensuring the safety and reliability of these indispensable components.

Beyond the chemical composition, the physical form of a primary explosive significantly impacts its sensitivity and overall performance. The particle size and crystal morphology influence the ease of initiation, with finer particles typically exhibiting greater sensitivity. Furthermore, the method of preparation and the presence of impurities can profoundly affect the stability and reliability of the explosive. Careful control over these parameters during manufacturing is crucial for ensuring consistent performance and mitigating the risk of accidental detonation. The precise control over these physical factors is often overlooked, despite their crucial contributions to the overall safety and effectiveness of the final product.

The applications of primary explosives span various sectors. In the mining and demolition industries, they serve as reliable initiators for larger charges of secondary explosives, ensuring controlled and efficient blasting operations. In the defense sector, primary explosives are vital components of ammunition, used to initiate the detonation of the main explosive charge in projectiles and warheads. Their ability to reliably initiate detonation under diverse conditions makes them indispensable in these applications.

Despite the diverse applications, the safety aspects of handling and storing primary explosives remain paramount and demand a high level of technical expertise.

The future of primary explosive technology points towards a shift toward greater safety and enhanced performance. Research into novel chemical compounds focuses on developing materials with reduced sensitivity, while retaining or even improving their detonation capabilities. This includes the exploration of environmentally benign alternatives to replace some of the older, toxic compounds. These advances will continue to drive improvements in safety procedures, manufacturing techniques, and the overall reliability of primary explosives in various applications. This focus on safety and sustainability will be critical in ensuring the continued responsible utilization of these powerful and essential materials. Ultimately, progress in this field requires a careful balance between maximizing explosive power and minimizing the risks associated with their production and use. It is a fascinating field with ongoing research that will influence a number of fields for years to come.

4.2 Secondary Explosives

Secondary explosives, unlike their primary counterparts, require a significant external stimulus—a detonator—to initiate their explosive decomposition. This characteristic makes them safer to handle and transport, contributing to their widespread use in military and industrial applications. Their relative insensitivity, however, doesn't diminish their immense power; a small quantity of a secondary explosive can generate a devastating blast wave.

The key distinction between primary and secondary explosives lies in their sensitivity. Primary explosives, like lead azide or mercury fulminate, detonate readily upon exposure to heat, friction, or impact. Secondary explosives, on the other hand, demand a much stronger stimulus. This difference fundamentally shapes their applications; primaries are predominantly used as detonators, initiating the explosion of the much less sensitive secondary explosive.

Several chemical families constitute the vast landscape of secondary explosives. Nitramines, for example, represent a significant class, with compounds such as RDX (cyclotrimethylenetrinitramine) and HMX (cyclotetramethylenetetranitramine) holding prominent positions. Their high energy density and relatively low sensitivity make them ideal for a variety of applications, from military munitions to demolition charges. RDX, in particular, is frequently encountered in plastic explosives due to its powerful detonation and ability to be molded into various forms.

Nitroaromatics, another important group, include well-known explosives like TNT (trinitrotoluene). While less powerful than nitramines, TNT's insensitivity, stability, and ease of melting make it a preferred choice for filling shells and bombs, where its relatively low sensitivity reduces the risk of accidental detonation during handling and transportation. The melting point of TNT allows for easy casting into various shapes and sizes, a crucial property for its widespread use in military ordnance.

Beyond nitramines and nitroaromatics, other secondary explosives exist, each with unique properties tailored to specific applications. For instance, PETN (pentaerythritol tetranitrate) is known for its high velocity of detonation, making it suitable for specialized applications requiring high brisance. This high velocity of detonation makes it particularly effective in shaping charges, used for cutting and forming metal, a vital tool in many industrial processes and military applications.

The synthesis of secondary explosives is a complex process, requiring precise control over reaction conditions to ensure purity and safety. Typically involving nitration reactions, these syntheses often involve handling hazardous chemicals and necessitate stringent safety protocols. Impurities introduced during the synthesis process can drastically alter the explosive's performance and sensitivity, potentially leading to unpredictable and dangerous outcomes. The precise control and highly specialized knowledge required highlight the importance of expertise and rigorous adherence to safety procedures.

The sensitivity of a secondary explosive is influenced by numerous factors, including its crystalline structure, particle size, and the presence of impurities. Smaller particle sizes generally increase sensitivity, making handling and storage more hazardous. The crystalline structure can significantly influence the material's impact sensitivity. Impurities, even in trace amounts, can act as catalysts, accelerating decomposition and increasing the risk of accidental detonation. Understanding these factors is crucial for safe handling, storage, and transportation.

Furthermore, the stability of secondary explosives over time is paramount. Factors such as temperature, humidity, and exposure to UV light can affect their stability, leading to potential degradation and increased sensitivity. Proper storage and handling protocols are crucial to ensure that these explosives remain stable and safe for their intended purpose, preventing unintended consequences due to prolonged exposure to environmental factors or aging. Rigorous quality control measures are essential throughout the entire lifecycle of these materials.

The environmental impact of secondary explosives is also a concern. The manufacture and use of these compounds can result in the release of toxic substances into the environment, demanding careful consideration of disposal and remediation strategies. The chemical constituents of the compounds can impact soil and water quality, and potentially threaten human and animal health. Therefore, minimizing the environmental impact and developing environmentally friendly alternatives are areas of ongoing research and development within the field.

Finally, understanding the specific characteristics of different secondary explosives is vital for selecting the most appropriate compound for a given application. Choosing a less-sensitive explosive may be critical in minimizing risks, while high energy density may be essential for particular military applications, and ease of handling is paramount for certain industrial purposes. This intricate interplay of properties necessitates a deep understanding of the materials' chemistry and behavior. The proper selection of secondary explosives involves carefully evaluating their properties and suitability within specific application parameters, reflecting a balance between power and safety.

5 Propellants and Pyrotechnics

Solid propellants form the backbone of many rocket and missile systems. Their chemistry is complex, involving a carefully balanced mixture of fuel, oxidizer, and often a binder to hold everything together. Common oxidizers include ammonium perchlorate (AP) or ammonium nitrate (AN), providing the oxygen necessary for combustion. Fuels can range from simple polymers like hydroxyl-terminated polybutadiene (HTPB) to more sophisticated energetic binders that contribute to overall performance. The choice of ingredients significantly impacts the propellant's burn rate, energy output, and mechanical properties. Careful consideration must be given to achieving a desired balance between these often-conflicting factors.

Liquid propellant systems, in contrast to their solid counterparts, offer greater control over the combustion process. This allows for improved thrust vectoring and engine throttling capabilities, making them particularly desirable for space launch vehicles and advanced maneuvering systems. Typical liquid propellants consist of separate fuel and oxidizer tanks, which are mixed and ignited only when necessary. Common examples include kerosene/liquid oxygen (LOX) combinations, which provide high energy density, and hypergolic propellants that ignite spontaneously upon contact. The latter category, while offering ease of ignition, often involves highly toxic and corrosive materials, demanding stringent safety precautions during handling and storage. Their high reactivity necessitates careful design considerations of the storage and delivery systems.

Pyrotechnics, while sharing some chemical similarities with propellants, differ fundamentally in their application. Their primary purpose is not propulsion but rather the generation of heat, light, smoke, or sound, and sometimes a combination thereof. This diverse range of effects stems from the careful selection and mixing of various components. These can include oxidizers like potassium perchlorate or potassium nitrate, fuels such as aluminum powder or magnesium, and colorants and binders to influence the resulting spectacle. Safety is paramount, especially given the potential for unintended ignition or explosion. Careful consideration must be given to formulation, particle size, and mixing techniques to avoid unintentional hazards. A detailed understanding of combustion chemistry is vital for the design of both safe and effective pyrotechnic devices. The precise control over the energetic materials provides a wide array of possibilities.

The design and application of propellants and pyrotechnics often require a fine balance between energy output, safety, and environmental impact. This entails a deep understanding of combustion thermodynamics, kinetics, and the physical properties of the constituent materials. Furthermore, the sensitivity of these materials to external stimuli, such as shock, friction, or electrostatic discharge, dictates strict safety protocols throughout their handling, storage, and deployment. For example, specific safety procedures must be developed during the manufacturing and handling of energetic materials, like the use of protective equipment and carefully planned process protocols. A systematic approach to risk assessment and mitigation is crucial in mitigating potential hazards associated with their use. It is important to regularly review and update safety protocols to mitigate any arising risks.

Optimization of propellant and pyrotechnic formulations is an ongoing process involving both empirical experimentation and advanced computational modeling techniques. This is particularly crucial in the quest for higher performance, improved safety, and reduced environmental impact. Novel materials and innovative approaches continuously emerge, pushing the boundaries of what's achievable in this rapidly evolving field. Computational chemistry and materials science play an increasingly important role in predicting the performance of new compounds and in designing safer formulations before actual physical testing. This computational approach is important because it is able to provide rapid evaluation and predictions before undergoing costly and time-consuming physical experiments. This combination of advanced modeling and testing provides a very powerful technique for both propellant and pyrotechnic development.

The environmental implications of energetic materials cannot be overlooked. Many traditional formulations contain substances that are toxic, polluting, or both. Therefore, a significant research effort is focused on developing greener, more environmentally benign alternatives, replacing hazardous components with less toxic substitutes. The increased focus on environmental concerns is driving the development of new and safer energetic materials. New formulations are focused on the use of alternative chemicals which are less harmful, and that result in reduced pollution and smaller environmental impacts. It's critical that the development of these materials follows all environmental regulations and environmental impact statements.

5.1 Solid Propellant Chemistry

Solid propellants, unlike their liquid counterparts, offer advantages in storage and handling due to their inherent stability. Their formulation, however, demands a precise understanding of chemical interactions and the resulting energetic output. A common base for many solid propellants is ammonium perchlorate (AP), a powerful oxidizer providing the necessary oxygen for combustion. This is often combined with a fuel, such as aluminum powder, which significantly enhances the energetic output and contributes to the overall performance. The precise ratio of oxidizer to fuel is crucial, meticulously determined through extensive testing to optimize burn rate and energy density. Additives are frequently incorporated to improve processing characteristics, mechanical properties, or even to modify the burning rate profile - for instance, the addition of certain binders affects the propellant's physical consistency and its reaction speed.

The binding agent is essential; it holds the solid propellant together, maintaining its structural integrity. Common choices include polybutadiene acrylonitrile (PBAN) or hydroxyl-terminated polybutadiene (HTPB), synthetic polymers that provide both the structural support and the fuel source, although their contribution to the overall energy release is generally less significant than AP or aluminum. These binders influence a variety of factors, including the propellant's sensitivity to shock and friction, its mechanical strength, and its overall longevity. Proper selection and implementation significantly impact the overall performance and safety of the final product. In essence, the binder functions as a structural matrix, influencing not only the physical properties but also the burning characteristics of the complete formulation.

Beyond the major components, several additives play crucial roles in optimizing propellant performance. Burning rate modifiers, often metallic salts or organic compounds, adjust the speed at which the propellant burns. This is essential for controlling thrust in rocket applications, where precise and predictable burn rates are paramount for mission success. Stabilizers, meanwhile, extend the propellant's shelf life by preventing unwanted decomposition or reactions that might compromise its performance or safety. These stabilizers counteract the inherent instability of some propellant components, thus ensuring reliability over longer periods of storage. Understanding their function and interaction with the main components is vital for designing dependable propellants.

The chemical reactions within a solid propellant are complex, involving multiple simultaneous processes. The initial ignition step initiates a chain reaction, causing the decomposition of the oxidizer (AP) and the fuel (aluminum, binder). This decomposition releases oxygen and gaseous fuel, which subsequently react at a controlled rate in the combustion zone. This combustion process generates high temperatures and produces enormous quantities of gases, which in turn generate the thrust. The complexity of the reactions makes modelling propellant behavior challenging. However, these models are crucial for performance predictions and safety assessments, allowing for the design of more efficient and predictable systems.

Designing a solid propellant is not just about selecting appropriate components; it also involves understanding their interactions. Compatibility is a critical consideration; the chosen ingredients must be stable together over their expected storage lifespan, resisting any undesirable reactions which might affect the performance and, crucially, the safety of the system. The manufacturing process significantly impacts the final product's properties, with consistent mixing and curing conditions crucial to ensuring homogeneity and performance consistency. Any inconsistencies in the manufacturing process can lead to variations in burn rate, pressure, and other critical aspects of the final propellant's performance. Careful attention must therefore be given to quality control measures at every stage of manufacturing.

Moreover, the environmental impact of solid propellants must be considered. The combustion process produces various byproducts, some of which can be harmful. The use of AP, while effective as an oxidizer, creates environmentally unfriendly byproducts. Current research focuses on the development of environmentally benign formulations, reducing reliance on hazardous materials. Alternative oxidizers and fuels are actively being explored, emphasizing both performance and environmental responsibility. This trend aligns with the growing global emphasis on environmentally sound practices within all sectors of industry, and this extends to the design and manufacture of energetic materials.

Finally, the safety aspects of handling and storing solid propellants are of paramount importance. They are inherently energetic and must be treated with appropriate care. Procedures for safe manufacturing, handling, and disposal are established and rigidly enforced to minimize the risk of accidental ignition or detonation. Stringent quality control measures are essential to guarantee consistent and predictable performance whilst reducing the possibility of failures. Research into safer formulations, with reduced sensitivity to impact and friction, is ongoing, with a focus on creating safer and more predictable propellants. This area of research continues to be a crucial aspect of the development of propellants and related energetic materials.

5.2 Liquid Propellant Systems

Liquid propellants offer a unique set of advantages and challenges compared to their solid counterparts. Their fluidity allows for precise control over the thrust produced by a rocket engine, a crucial factor in maneuvering and achieving specific trajectories. This control is achieved through regulating the flow rates of the fuel and oxidizer components, a capability not easily replicated with solid propellants. The ability to restart and throttle engines using liquid propellants also provides flexibility in mission profiles, allowing for mid-flight course corrections or adjustments in propulsion needs. However, this increased complexity comes at a cost; liquid propellant systems are often more intricate and require extensive and precise engineering for reliable operation. Storage and handling also introduce significant safety concerns, demanding robust safety protocols and sophisticated engineering solutions. Furthermore, the need for complex pumping and pressurization systems increases the overall weight and complexity of the system.

The selection of a propellant combination is dictated by several factors including specific impulse, density, storage characteristics, and toxicity. Specific impulse, a measure of a propellant's efficiency, is a critical performance parameter. Propellants with high specific impulse deliver more thrust per unit of propellant mass, leading to greater payload capacity or extended mission durations. Density plays a critical role in determining the overall volume and weight of the propellant tanks, thus influencing the overall design and performance of the rocket. A propellant with high density is preferable, as it allows for a smaller and lighter vehicle. Storage characteristics, like volatility and sensitivity to temperature changes, are vital for safe handling and reliable performance. Propellants susceptible to decomposition or explosive behavior pose significant challenges in terms of storage and transportation. Toxicity considerations are paramount; propellants emitting harmful byproducts during combustion demand stringent handling procedures and careful environmental impact assessments, necessitating elaborate safety measures and post-combustion mitigation strategies.

Among the common liquid propellant combinations are those employing kerosene or RP-1 (refined kerosene) as the fuel and liquid oxygen (LOX) as the oxidizer. This combination has been a mainstay in rocketry for decades, offering a balance between performance, cost, and relative ease of handling, although challenges remain in LOX's cryogenic nature. LOX requires sophisticated insulation and management to prevent boil-off, particularly in long-duration missions. Another popular combination involves using hydrazine (N2H4) or one of its derivatives as a fuel, often paired with dinitrogen tetroxide (N2O4) as the oxidizer. These hypergolic propellants ignite spontaneously upon contact, eliminating the need for an ignition system, simplifying the engine design, though presenting significant handling challenges due to the toxicity of these compounds. Their hypergolic nature reduces system complexity. However, the inherent toxicity necessitates strict handling protocols and the implementation of comprehensive safety measures. The selection between these combinations and others is heavily influenced by the specific mission requirements, balancing performance needs against safety, cost, and environmental considerations.

The design of liquid propellant tanks involves careful consideration of material selection, tank shape, and pressure management. Tank materials must withstand extreme temperatures and pressures and be compatible with the chosen propellants, preventing corrosion or degradation. Tank shape and size influence the overall spacecraft design, optimizing for volume and minimizing weight. Pressure management is critical for maintaining consistent propellant flow rates and preventing excessive pressure buildup. These considerations often necessitate the use of specialized alloys, advanced composite materials, and intricate pressure regulation systems, leading to complex designs and manufacturing processes. Furthermore, cryogenic propellant tanks, used for storing substances like liquid hydrogen and LOX, require advanced insulation technologies to minimize boil-off and maintain the propellant's desired temperature. Precise design and manufacturing tolerances are essential to prevent leaks or other failures that can compromise mission success, leading to rigorous quality control and thorough testing regimes.

Finally, the safe and efficient operation of liquid propellant systems demands a comprehensive approach to safety and control. This necessitates meticulously designed systems to prevent leaks, spills, and unintended ignitions, demanding rigorous safety protocols throughout the entire lifecycle. The implementation of redundant systems and multiple layers of safety mechanisms is crucial in mitigating the risks associated with the inherent hazards of working with volatile and potentially toxic substances. Continuous monitoring and careful control of temperature, pressure, and flow rates are essential for maintaining stable operation and preventing catastrophic failures. Comprehensive training for personnel involved in handling and operating these systems is of paramount importance, emphasizing procedures for safe handling, emergency response, and maintenance. These intricate systems require a level of expertise and planning that goes beyond simple mechanical considerations, demanding a deep understanding of both engineering principles and safety considerations.

5.3 Pyrotechnic Compositions

Pyrotechnic compositions represent a fascinating blend of chemistry and artistry, yielding vibrant displays and crucial functionalities. These mixtures, unlike high explosives designed for maximum brisance, are carefully formulated to produce specific effects, such as light, color, smoke, or sound, at controlled rates and intensities. The key to understanding their behavior lies in the intricate interplay between the oxidizer, fuel, and any additives present. A typical composition consists of an oxidizer, providing the oxygen necessary for combustion; a fuel, which burns readily in the presence of oxygen, generating heat and light; and binders, which hold the mixture together, ensuring its structural integrity.

The choice of oxidizer is pivotal, dictating the overall performance and characteristics of the pyrotechnic mixture. Common oxidizers include nitrates like potassium nitrate (KNO), perchlorates such as ammonium perchlorate (NHClO), and even metal oxides like barium peroxide (BaO). Each offers unique properties; potassium nitrate, for instance, is known for its relatively low cost and relatively low sensitivity, making it suitable for many consumer fireworks. Ammonium perchlorate, on the other hand, delivers higher energy but poses greater safety considerations due to its higher sensitivity to impact and friction. Careful consideration of the oxidizer's characteristics—its oxygen balance, its thermal stability, its compatibility with other components—is crucial for a safe and effective pyrotechnic composition.

Fuels in pyrotechnic compositions are equally diverse. Metals such as aluminum and magnesium are frequently used for their high energy content and bright emissions. Aluminum produces a brilliant white light, while magnesium generates an intense white-blue flash. Organic fuels, like various carbohydrates or synthetic polymers, provide additional energy and can influence the burn rate and flame characteristics. The selection of the fuel strongly impacts the speed and efficiency of the combustion process, as well as the overall color and intensity of the light produced. The ratio of oxidizer to fuel is often finely tuned to achieve the desired effect, ensuring complete combustion without excessive smoke or residue.

Beyond the primary oxidizer and fuel components, a wide array of additives can be incorporated to tailor the pyrotechnic effect. These additives modify the color, intensity, duration, and other specific properties of the combustion. Metal salts, for example, play a critical role in generating color: strontium salts produce red, calcium salts orange, barium salts green, copper salts blue, and sodium salts yellow. These salts vaporize during combustion, emitting light at specific wavelengths that dictate the color perceived by the observer. The concentration of these metallic salts directly influences the vibrancy and intensity of the resultant color.

Binders are an often-overlooked, yet essential, component. They serve to hold the oxidizer, fuel, and other additives together, forming a cohesive mixture that maintains its structural integrity during preparation, storage, and ignition. Common binders include starches, dextrins, and polymers. The choice of binder can impact the burn rate and the mechanical properties of the final composition. The correct selection of binder is vital in ensuring consistent ignition and burn characteristics, vital for safety and reliability.

The production of pyrotechnic compositions necessitates careful handling and precise mixing techniques. Thorough blending is crucial to ensure a uniform mixture, which minimizes the risk of localized hotspots and unpredictable combustion. The exact procedure varies based on the specific composition and the intended application. Safety precautions are paramount, including the use of appropriate personal protective equipment, working in well-ventilated areas, and avoiding any potential sources of ignition. The entire process requires not only chemical expertise but also a detailed understanding of safety protocols.

Finally, the performance of any pyrotechnic composition is directly tied to its sensitivity and stability. These attributes are influenced by various factors, including the particle size of the components, the specific oxidizer and fuel used, and the presence of any sensitizing agents. A thorough understanding of these factors is essential for designing safe and reliable compositions. Sensitivity testing and rigorous quality control measures are essential to ensure the intended behavior of the product and the safety of those involved in its manufacturing and use.

6 Insensitive Munitions

The drive towards safer handling and storage of energetic materials has led to significant advancements in insensitive munitions (IM) technology. This necessitates a departure from traditional high-energy density formulations, often prioritizing sensitivity reduction over peak performance. The challenge lies in finding a balance—maintaining sufficient explosive power while significantly mitigating the risk of accidental detonation. This requires a sophisticated understanding of the chemical and physical properties of the energetic materials, coupled with innovative design approaches.

Reduced sensitivity formulations are at the core of IM development. This involves incorporating various additives or modifying the crystal structure of the primary explosive to increase the energy required for initiation. Examples include the use of polymeric binders that encapsulate explosive crystals, thus isolating them from external stimuli. Other approaches involve incorporating desensitizing agents which physically or chemically interrupt the propagation of shock waves within the explosive charge. Each method carries its own set of trade-offs, and selecting the appropriate approach depends on the specific application and desired performance characteristics.

Designing insensitive munitions necessitates a holistic approach, considering not only the formulation but also the overall system design. This includes the physical packaging of the explosive, the use of safety features like fuses and initiators, and even the transportation and storage conditions. Proper design can minimize the propagation of shock waves generated either internally from inherent material defects, or externally, through impact or friction. Robust container designs, for instance, capable of withstanding significant external shock, form a vital part of this overall strategy. Understanding the potential failure modes of the entire system is critical for effective IM design.

Rigorous safety testing is paramount. A comprehensive suite of tests, tailored to the specific IM, is necessary to verify its performance and ensure its insensitivity. These tests often involve subjecting the munition to a range of stimuli, such as impact, friction, and electrostatic discharge, to assess its response. Standardized protocols, like those outlined by international organizations, are typically followed to ensure consistency and comparability across different IM designs and manufacturers. Data analysis from these tests is then used to optimize the design further, creating a feedback loop of refinement.

The field of insensitive munitions is constantly evolving. Current research explores novel energetic materials with inherently lower sensitivity, leveraging advancements in materials science and computational chemistry. This includes the investigation of new crystal structures or molecular modifications to optimize both energy density and safety. The development of more environmentally friendly energetic materials is also a growing area of interest, pushing toward more sustainable solutions for military and industrial applications. There's a continued focus on developing new testing methodologies to further refine safety criteria and ensure future IM designs meet increasingly stringent safety standards.

Beyond the development of new materials and sophisticated testing methods, there's a growing emphasis on predicting sensitivity using computational models. These models, while still under development, hold significant promise in accelerating the design process and reducing the reliance on extensive experimental testing. By simulating the behavior of energetic materials under various stimuli, researchers can identify promising candidates for IM formulations, streamlining the development cycle and reducing costs. Ultimately, the goal is to achieve a balance between performance and safety that surpasses existing standards, making these powerful compounds safer to handle and use.

6.1 Reduced Sensitivity Formulations

Developing insensitive munitions is paramount for minimizing accidental detonations and enhancing safety. This necessitates a shift towards reduced sensitivity formulations, which involve modifying the chemical composition or physical structure of energetic materials to raise their threshold for initiation. This isn't simply about making things safer; it's about refining our understanding of energetic material behavior to create more predictable and reliable systems.

One crucial approach involves incorporating chemical additives that act as energetic material desensitizers. These additives, often polymers or plasticizers, can physically interrupt the propagation of shockwaves, thereby preventing chain reactions that lead to detonation. Careful selection of these additives is crucial; their compatibility with the base energetic material and their impact on performance must be thoroughly evaluated. For example, certain polymers can act as binders, improving the mechanical strength of the formulation while simultaneously reducing sensitivity.

The use of energetic binders represents another promising avenue. Traditional binders primarily serve to hold the energetic components together, but advanced binders can also participate directly in the energy release process, potentially improving overall performance while simultaneously decreasing the sensitivity to impact or friction. The choice of binder depends heavily on the specific energetic material and the desired properties of the final product. Understanding the interplay between binder chemistry, energetic material properties, and overall sensitivity is key. This requires a deep understanding of polymer science and its intersection with energetic material chemistry. Advanced characterization techniques are critical to evaluating the effectiveness of these modifications.

Beyond chemical modifications, physical approaches to desensitization also hold considerable potential. One such technique involves microencapsulation, which involves coating individual energetic particles with a protective layer. This layer can act as a physical barrier, preventing the initiation of a chain reaction, even under considerable stress. The choice of encapsulant material is critical, as it must provide sufficient protection without significantly hindering the performance of the energetic material. Optimizing the size and morphology of the encapsulated particles is another crucial factor affecting overall sensitivity.

Moreover, the careful control of particle size and morphology offers another route to reduce sensitivity. Smaller, more uniformly sized particles can often lead to more stable and less sensitive formulations. This is because smaller particles offer less surface area for initiation and are less susceptible to hot spot formation. Advanced manufacturing techniques, such as spray drying or nanotechnology, can be used to precisely control the particle size and morphology, allowing for fine-tuning of the final product's sensitivity characteristics.

The design of reduced sensitivity formulations is an iterative process, requiring careful consideration of several factors, including the energetic material's chemical properties, physical characteristics, and intended application. Rigorous safety testing is indispensable throughout the process to ensure that the modified formulations meet the desired safety standards while maintaining adequate performance. The development of new insensitive munitions is not a mere matter of tweaking existing compounds; it demands a thorough understanding of the complex interplay of chemistry, physics, and engineering principles.

Furthermore, the development of insensitive munitions demands a comprehensive approach, integrating materials science, chemical engineering, and advanced testing methodologies. This interdisciplinary approach is crucial to creating formulations that meet the stringent safety requirements while retaining sufficient performance characteristics for their intended application. The advancements in these fields have enabled the creation of increasingly sophisticated and safe energetic materials.

Ultimately, the transition to insensitive munitions requires not only the development of reduced sensitivity formulations but also a shift in mindset. This involves a renewed focus on safety, a commitment to rigorous testing, and the integration of advanced technologies. It demands not only expertise in chemistry and material science but also a deep understanding of the risks associated with energetic materials and the measures necessary to mitigate them. The ongoing research and development in this field are critical to ensuring the safe and responsible use of energetic materials in both military and civilian applications. Therefore, continuous innovation is paramount to keep pace with the constantly evolving demands for safer and more reliable energetic systems. The ongoing pursuit of improved safety should never compromise the necessary performance characteristics of the energetic material. The goal is to achieve a balance, ensuring both safety and effectiveness. The integration of advanced computational tools allows for a more efficient development process, accelerating the transition to a safer future.

6.2 Design Considerations

Designing insensitive munitions requires a multi-faceted approach, prioritizing safety without compromising performance. The fundamental challenge lies in mitigating the sensitivity of energetic materials to accidental initiation while maintaining their desired explosive power. This involves careful consideration of several key factors, starting with the chemical composition itself. The selection of less sensitive energetic materials, such as those with reduced impact or friction sensitivity, forms the cornerstone of insensitive munition design. This often entails utilizing less energetic compounds or incorporating additives that act as energy dampeners. Careful consideration must be given to the crystal structure, particle size, and morphology of the energetic material. Fine particles, for example, can be far more sensitive than coarser ones due to increased surface area. Controlling these factors can significantly influence the overall sensitivity of the formulation.

Beyond the chemical composition, the physical design of the munition itself plays a crucial role in safety. This includes the geometry of the explosive charge, the type of casing material, and the presence of any safety features such as inert barriers or confinement systems. Designing the munition to distribute or absorb any impact energy away from the explosive charge is critical. The use of protective layers or specialized coatings can further enhance safety by absorbing external shocks or limiting friction. The selection of the appropriate casing material requires consideration of mechanical properties like strength and resistance to deformation, alongside compatibility with the energetic materials and environmental factors such as temperature and humidity.

Furthermore, the manufacturing process significantly impacts the final product's sensitivity. Manufacturing inconsistencies can introduce defects that increase the risk of accidental detonation. Therefore, rigorous quality control measures are essential throughout the entire manufacturing process, including raw material selection, mixing techniques, and packaging procedures. Careful consideration should be given to preventing electrostatic discharge and ensuring consistent material properties during handling and transportation. Stringent environmental controls within the manufacturing facility are vital to minimize the risks associated with static electricity build-up, and humidity levels should be monitored and controlled to prevent unwanted changes in the material's properties.

The incorporation of safety features beyond the basic design is crucial for truly insensitive munitions. This might include the use of shock-absorbing materials to dampen impacts, thermal fuses to prevent premature detonation, or mechanical devices that inhibit the propagation of detonation. These safety features are often carefully integrated with the munitions' other design elements. For example, a protective case around the explosive might include thermal insulation, preventing excessive temperature increases that could lead to undesired ignition, alongside mechanical barriers to prevent the propagation of shockwaves. Careful testing and modeling are required to ensure these features are both effective and reliable.

Finally, thorough testing is paramount to validating the effectiveness of the design considerations implemented. A comprehensive testing program should incorporate a range of sensitivity tests, including impact, friction, and electrostatic discharge tests, and also evaluate the munition's performance under various environmental conditions. The results of these tests are essential in verifying the safety and reliability of the insensitive munition. They not only help confirm the safety levels of the design, but also provide valuable feedback for further improvements and optimization. This data-driven approach ensures that the final product meets the desired safety standards while still maintaining satisfactory performance capabilities. The overall goal remains a balance between safety and effectiveness. Therefore, every stage of the design process must carefully consider this delicate equilibrium.

6.3 Safety Testing Methods

Safety testing is paramount in handling energetic materials. Neglecting rigorous testing can lead to catastrophic accidents, emphasizing the absolute necessity of comprehensive evaluation before any application. These tests assess a compound's susceptibility to unintended initiation, providing crucial data for safe handling, storage, and transportation protocols. The selection of appropriate tests depends heavily on the material's intended use, its chemical composition, and the potential hazards associated with it. Impact sensitivity tests measure a material's resistance to initiation from mechanical shock. The most common methods involve dropping a weight onto a small sample of the compound. The height from which the weight must be dropped to cause initiation is a key parameter, providing an indication of the material's sensitivity to impact. Variations exist in hammer drop tests, utilizing different weights and sample configurations to provide a more complete picture of the material's response. A related test is the friction sensitivity test, measuring the ease of initiation by frictional forces. This is often conducted by rubbing a small amount of the compound against a rough surface under controlled conditions.

Friction sensitivity testing involves a standardized apparatus to evaluate the likelihood of initiation due to friction. The compound is subjected to controlled frictional forces, and the outcome is a critical assessment of its potential hazards. Results are quantified and compared to well-established standards within the field, determining the material's safety profile under frictional conditions. These tests help establish safe handling procedures, for example, specifying the types of containers and surfaces suitable for handling, storage, and transportation of the energetic material.

Thermal stability analysis is critical for determining a compound's susceptibility to decomposition or explosion under heat. Differential scanning calorimetry (DSC) and thermogravimetric analysis (TGA) are widely employed. DSC measures the heat flow associated with phase transitions and chemical reactions, providing insights into the material's thermal behavior. TGA, conversely, measures mass changes as a function of temperature, revealing decomposition patterns. Together, these techniques provide a comprehensive understanding of the material's thermal stability profile, including the onset temperature of decomposition and the energy released during this process. Data from these tests inform safe storage temperature ranges, preventing accidental decomposition or self-ignition.

Electrostatic discharge (ESD) sensitivity is another crucial safety parameter, especially for materials used in sensitive electronic environments. ESD tests simulate accidental electrical discharges to assess the likelihood of initiation. This involves subjecting the material to various levels of electrostatic discharge and observing the resultant response. Testing parameters include the voltage level, the duration of the discharge, and the distance between the electrodes and the sample. The results provide critical data that can help to design and implement ESD protection measures, minimizing the risk of accidental initiation.

Detonation velocity measurements provide insight into the compound's explosive power, an important characteristic for both safety and performance evaluation. Techniques involve employing sensors to measure the speed at which a detonation wave propagates through a column of the material. Data from such tests is critical in determining a material's suitability for its intended application and understanding the potential hazards associated with its detonation. This is especially important for larger-scale applications where the consequences of detonation are significant. This data directly informs the safe design of explosive devices and their containment systems.

Water sensitivity is crucial to assess the material's stability in various environments. The tests commonly involve immersing a small amount of the compound in water under controlled conditions, observing its behavior, and quantifying the extent of any changes. The duration of exposure, water temperature, and subsequent changes in the sample's properties are carefully documented. The impact of water sensitivity is vast, affecting material storage, transportation, and environmental considerations. Understanding the reactions with water can improve storage solutions and allow for the accurate assessment of potential environmental impacts.

While these methods offer vital insights into the safety profile of energetic materials, no single test provides a complete picture. Comprehensive safety assessments require a combination of these tests, tailored to the specific material and its intended use. The synergistic use of multiple testing techniques allows researchers and manufacturers to fully understand the material's behavior under different conditions, leading to safer handling, storage, transportation, and ultimately, more reliable applications. Furthermore, continuous improvement and standardization in safety testing protocols ensure a higher level of safety in the field. Advancements in these areas are crucial for the responsible development and implementation of energetic materials.

6.4 Future Directions

The pursuit of safer, more efficient, and environmentally responsible energetic materials remains a paramount challenge. Current research focuses heavily on the development of novel insensitive munitions, moving beyond simply reducing sensitivity to exploring fundamentally new approaches to energetic material design. This necessitates a deeper understanding of the complex interplay between molecular structure, chemical reactivity, and macroscopic performance. Advanced computational modeling, incorporating density functional theory (DFT) and reactive molecular dynamics (RMD), will play a pivotal role in accelerating the discovery and optimization of these next-generation materials.

Minimizing the environmental impact of energetic materials manufacturing and use is becoming increasingly crucial. The development of green energetic materials, synthesized from readily available and less toxic precursors, with biodegradable byproducts, presents a significant opportunity. This requires moving away from traditional, environmentally harmful components and exploring alternatives such as those based on nitrogen-rich heterocycles or ionic liquids. Life-cycle assessments of the entire manufacturing and disposal process will be vital in truly evaluating the 'greenness' of any new compound.

Beyond the environmental considerations, the need for increased performance in specific applications drives continued research. This includes the development of high-energy-density materials for advanced propulsion systems, seeking to achieve higher specific impulses while maintaining compatibility with existing infrastructure. Similarly, enhanced performance in specialized pyrotechnic applications, such as those requiring tailored spectral emissions or precise timing, necessitates the exploration of novel chemical formulations and sophisticated ignition mechanisms.

A critical area of future development involves the integration of advanced diagnostics and characterization techniques. Real-time monitoring of energetic material performance under extreme conditions requires sophisticated instrumentation and data analysis methods. This includes advanced imaging techniques capable of resolving micro-scale events during detonation, enabling a more detailed understanding of the underlying processes. Improvements in sensor technology will enhance our ability to predict and control the behavior of energetic materials, leading to safer and more reliable systems.

The field of energetic materials research is poised for significant advancements through interdisciplinary collaborations. By combining the expertise of chemists, physicists, engineers, and materials scientists, we can accelerate progress towards safer, more efficient, and sustainable energetic materials. These collaborative efforts will foster the development of new theoretical models, experimental techniques, and computational tools that will ultimately reshape the landscape of energetic materials research and development. This multi-faceted approach is essential in addressing the complex challenges associated with energetic materials, ensuring responsible innovation in a rapidly evolving technological landscape. The integration of artificial intelligence and machine learning tools promises to further accelerate this progress, allowing for the rapid screening and optimization of countless potential candidates. However, a parallel effort is crucial to develop robust safety protocols and risk assessments that are integral to the responsible advancement of this field. The future of energetic materials relies on a nuanced approach that balances the pursuit of advanced capabilities with an unwavering commitment to safety and environmental stewardship. Only through such a synergistic combination of advanced technologies, collaborative research and a heightened awareness of the potential risks and rewards can the field of energetic materials continue to progress in a manner that is both impactful and responsible.

Developing advanced manufacturing techniques that can produce energetic materials with greater precision and control is paramount. This includes exploring novel approaches such as 3D printing of energetic composites and the utilization of advanced material processing techniques to optimize the structure and properties of these complex systems. The development of more precise and controllable methods for manufacturing will lead to enhanced consistency in the performance characteristics of energetic materials, leading to more reliable and predictable systems. Furthermore, exploration into the design of self-healing energetic materials will provide enhanced reliability and safety, preventing catastrophic failures through an inherent capacity for self-repair. Such systems require sophisticated material design, and an understanding of their response to different failure scenarios.

Moreover, the exploration of novel energetic materials based on previously unexploited chemical systems will continue to expand the range of available options. This necessitates a thorough understanding of the fundamental chemical principles governing their behavior, encompassing both their synthesis, characterization, and properties. The future of energetic materials will rely heavily on the ability to discover and develop new molecules with superior performance characteristics. These molecules may require advanced synthesis methods, and their properties must be meticulously investigated to ensure safe and reliable application. Through combining experimental efforts with computational modeling, future research will undoubtedly uncover new avenues towards the creation of innovative and improved energetic materials. This holistic approach will enable us to tailor the design of these compounds to specific needs, enabling breakthroughs in a multitude of fields. This holistic approach of experimental research coupled with advanced computational methods promises to unlock revolutionary advances in the field, paving the way for new generations of safer and more efficient energetic materials.

6.5 Emerging Technologies

The pursuit of safer and more efficient energetic materials constantly drives innovation. One promising area lies in the development of novel energetic materials based on less sensitive, more environmentally friendly compounds. This involves exploring alternative chemical structures that minimize the risk of accidental detonation while maintaining, or even enhancing, performance characteristics. Research focuses on incorporating less hazardous elements and reducing the reliance on traditional, environmentally problematic components. This shift requires meticulous synthesis and characterization techniques, coupled with advanced computational modeling to predict performance and stability before undertaking large-scale synthesis.

Current research extensively utilizes advanced computational methods to design and predict the properties of energetic materials. Density functional theory (DFT) calculations, molecular dynamics simulations, and reactive molecular dynamics (ReaxFF) are invaluable tools. They enable researchers to virtually screen a vast array of potential candidates, identifying promising compounds before entering the laboratory. This computational screening process significantly reduces the time and cost associated with traditional experimental methods. Furthermore, it allows the exploration of chemical spaces that might be otherwise inaccessible or too hazardous to explore experimentally. This theoretical approach becomes essential in the discovery and refinement of new insensitive munitions.

Beyond computational design, the synthesis of energetic materials is undergoing a transformation. Microfluidic techniques, for example, offer precise control over reaction conditions, leading to the creation of highly homogeneous and well-defined energetic materials with improved performance and reduced sensitivity. The ability to precisely control particle size and morphology through microfluidic synthesis is particularly valuable in enhancing safety and performance. This level of control minimizes the formation of hotspots—tiny regions of high energy density that can trigger premature detonation. The synthesis of nanostructured energetic materials also shows promise, potentially leading to superior energy density and controlled burn rates.

Another area of active research is the incorporation of energetic nanoparticles within a matrix of less sensitive materials. This composite approach aims to leverage the high energy density of nanoparticles while mitigating their inherent sensitivity. The design of these composites is complex, requiring careful consideration of particle size, distribution, and interaction with the matrix material. The goal is to achieve a synergistic effect, where the overall performance of the composite surpasses the individual components. This includes improving burn rates, increasing overall energy density, and decreasing sensitivity. The precise control over the morphology of the incorporated energetic nanoparticles is vital in achieving the desired balance between energy and safety.

Moreover, the development of "green" energetic materials presents a crucial challenge and opportunity. The environmental impact of traditional energetic materials is substantial, and the search for more environmentally benign alternatives is critical. Researchers are exploring biodegradable and less toxic compounds, focusing on elements with lower environmental toxicity. This pursuit necessitates a thorough life-cycle assessment of the entire production, use, and disposal processes. The focus is on minimizing both the immediate and long-term environmental consequences associated with the production and application of these critical materials.

This exploration of greener alternatives often involves utilizing bio-based materials or incorporating readily available and renewable resources into energetic material formulations. This integration of bio-inspired designs aims to reduce the environmental footprint while minimizing the reliance on scarce or hazardous raw materials. The long-term goal is to develop energetic materials that are both high-performing and sustainable, balancing technological advancement with environmental responsibility. Such methods are essential in minimizing long-term environmental damage, as the use and disposal of energetic materials often leads to detrimental ecological impacts.

The field of energetic materials is constantly evolving, driven by the demands for enhanced performance and increased safety. The convergence of advanced computational design, innovative synthesis techniques, and the pursuit of environmentally benign materials is leading to the development of new generations of energetic compounds. These new technologies promise improved performance, increased safety, and reduced environmental impact, revolutionizing various applications ranging from defense and aerospace to industrial processes. The advancement in this field ensures a continued progress in the future, ensuring a balanced approach between technological advancement and ecological consciousness. The challenges remain significant, but the ongoing research and development efforts pave the way for a safer and more sustainable future in energetic materials technology.

7 Future Trends in Energetic Materials

Computational design is revolutionizing the field. Sophisticated software and advanced algorithms allow researchers to predict the properties of energetic materials with unprecedented accuracy, significantly reducing the need for costly and time-consuming experimental synthesis and testing. This predictive power allows for the rapid screening of thousands of potential candidates, identifying those most likely to possess desirable characteristics such as high energy density, low sensitivity, and improved stability. We're moving beyond trial and error towards a more rational and efficient design process. The integration of machine learning techniques further enhances this capability, enabling the identification of complex structure-property relationships that might otherwise remain hidden. This transition towards computational design is not simply about efficiency; it also opens doors to entirely new classes of energetic materials that would be difficult, if not impossible, to discover through traditional experimental methods.

The environmental impact of current energetic materials is a pressing concern. Traditional energetic materials often contain toxic heavy metals or produce harmful byproducts upon detonation, posing significant risks to both human health and the environment. The development of "green" energetic materials is therefore a high priority, focusing on compounds made from less toxic ingredients and producing fewer harmful emissions. This involves exploring alternative oxidizers and fuels, such as those based on nitrogen-rich heterocycles or metal-free compounds. Bio-based energetic materials, derived from renewable resources, represent another promising area, offering a sustainable alternative to traditional petroleum-based formulations. Researchers are actively investigating various green synthesis routes, aiming to minimize waste generation and energy consumption during the manufacturing process. This shift toward greener options is not just an ethical imperative; it also represents a significant opportunity to enhance the sustainability and long-term viability of the energetic materials industry. Furthermore, the lifecycle assessment of green energetic materials provides a quantitative measure of their environmental impact, facilitating informed decision-making throughout the design, production, and disposal phases.

A key challenge in green energetic material research is balancing performance with environmental friendliness. Often, environmentally benign compounds have lower energy densities compared to their conventional counterparts. Therefore, a significant focus is on developing innovative strategies to enhance the performance of green energetic materials without compromising their environmental benefits. This often involves manipulating the crystal structure or incorporating additives to optimize energy release and reduce sensitivity. Researchers are exploring novel approaches such as the use of energetic cocrystals, which combine two or more energetic components to improve performance and stability while retaining environmental benefits. The use of nano-sized energetic materials is another exciting development, offering the potential to enhance energy density and detonation velocity. However, handling and safety concerns associated with nano-materials require careful consideration. Overall, finding the optimal balance between performance and environmental compatibility requires a multidisciplinary approach, integrating chemistry, engineering, and environmental science expertise.

Beyond green chemistry, another major trend is the pursuit of improved safety. Insensitive munitions (IM) technologies continue to evolve, focusing on reducing the sensitivity of energetic materials to accidental detonation. This involves developing new formulations with decreased sensitivity to impact, friction, and electrostatic discharge. Innovative approaches include the use of energetic binders, polymeric coatings, and nano-confinement techniques to encapsulate sensitive components. Simultaneously, advanced characterization techniques are being developed to better understand the mechanisms of initiation and propagation in energetic materials, providing valuable insights for the design of safer formulations. This includes in-depth studies of shock sensitivity and the development of more accurate and reliable safety testing protocols. The overarching goal is to create energetic materials that are both potent and safe, ensuring the protection of personnel and reducing the risk of accidental explosions.

The convergence of different research areas is further shaping the future. For instance, the integration of computational design with green chemistry approaches enables the efficient discovery of environmentally benign energetic materials with enhanced performance. This synergistic combination accelerates the development cycle, enabling the rapid identification and optimization of promising candidates. The advancements in nanotechnology also contribute to these trends, opening new possibilities in terms of energy density, sensitivity, and performance enhancement. However, the unique challenges associated with nano-energetic materials, particularly in terms of safety and handling, necessitate rigorous research and careful consideration. Furthermore, the continued development of advanced characterization techniques enables a deeper understanding of the relationships between the molecular structure, properties, and performance of energetic materials, accelerating innovation in the field. This multi-faceted approach ensures that future energetic materials will be not only more powerful and efficient but also safer and environmentally sustainable.

7.1 Computational Design

Computational design offers a transformative approach to energetic materials research, dramatically accelerating the discovery and optimization of novel compounds. Instead of relying solely on trial-and-error experimentation, which is time-consuming and resource-intensive, computational methods leverage the power of quantum mechanics and advanced algorithms to predict the properties of molecules before they are ever synthesized. This predictive capability allows researchers to virtually screen thousands of potential candidates, identifying promising leads that warrant further investigation in the laboratory. This significantly reduces the need for extensive, and potentially hazardous, experimental work.

The core of computational design for energetic materials lies in sophisticated software packages capable of accurately modeling molecular structures and their behavior under various conditions. Density functional theory (DFT), a particularly powerful quantum mechanical method, allows for the calculation of key properties such as heat of formation, detonation velocity, and sensitivity. These calculations provide vital insights into the thermodynamic stability and energetic performance of a compound, guiding the design of molecules with enhanced properties. Furthermore, molecular dynamics simulations can model the response of materials to external stimuli, such as shock or impact, enabling the prediction of their sensitivity and safety characteristics. The integration of these methods within a high-throughput computational workflow allows for rapid assessment of a vast chemical space, a task impossible through traditional experimental approaches.

One of the most significant advantages of computational design is its ability to explore chemical space far beyond the limitations of experimental synthesis. Traditional approaches often focus on modifications of existing compounds, limiting the scope of discovery. Computational methods, however, can explore the properties of entirely novel structures and compositions, leading to the identification of unexpected energetic materials with superior performance characteristics. For example, computational studies have already identified promising candidates amongst novel nitrogen-rich heterocycles and metal-organic frameworks, structures not readily accessible through traditional synthesis routes. This allows for the design of materials with tailored properties, such as increased energy density, improved safety, or reduced environmental impact.

However, it's crucial to acknowledge the limitations of computational design. The accuracy of computational predictions relies heavily on the quality of the theoretical models and the computational resources available. While DFT and other methods provide increasingly accurate results, approximations are still necessary, and the results must be interpreted carefully. Experimental validation remains an essential step in the development process, confirming the computational predictions and characterizing the physical properties of synthesized compounds. Furthermore, computational studies must be complemented by careful consideration of synthesis feasibility and scalability, ensuring that computationally promising compounds can be effectively produced in practical quantities. It's a collaborative dance between theory and experiment.

The future of energetic materials development is inextricably linked with advanced computational design. As computing power increases and theoretical methods improve in accuracy, computational design will play an ever-larger role in shaping the field. The ability to predict the properties of energetic materials before synthesis offers not only significant time and cost savings but also opens up new avenues for innovation, potentially leading to the discovery of superior energetic materials with enhanced safety and reduced environmental impact. The integration of machine learning techniques with quantum chemical methods further promises to accelerate the pace of discovery, allowing for the exploration of even larger chemical spaces and the identification of previously unforeseen energetic material candidates. By carefully considering the strengths and limitations of computational methods and integrating them with experimental validation, researchers can unlock the full potential of computational design to drive the future of energetic materials. The convergence of cutting-edge computational tools and traditional experimental approaches will ultimately ensure the development of safer, more efficient, and more environmentally friendly energetic materials for both military and civilian applications, a pivotal advancement for the field. This synergistic approach will continue refining and optimizing our understanding of these powerful compounds.

7.2 Green Energetic Materials

The pursuit of greener energetic materials is paramount, driven by escalating environmental concerns and stringent regulations. Traditional energetic materials often leave behind toxic residues, posing significant threats to ecosystems and human health. This necessitates a shift towards sustainable alternatives that minimize environmental impact while maintaining, or even improving, performance characteristics. The challenge lies in discovering and developing compounds that are both powerful and environmentally benign, a delicate balance requiring innovative approaches in material science and chemical engineering.

One promising avenue lies in exploring alternative oxidizers. Perchlorates, a staple in many energetic formulations, are notorious for their environmental persistence and toxicity. Replacing them with less harmful oxidizers, such as nitrates or even oxygen-rich organic molecules, offers a pathway to reduced environmental burden. This substitution requires careful consideration of the energetic material's overall performance, as the replacement oxidizer might exhibit lower oxidizing power, necessitating adjustments in the fuel component or overall formulation. Extensive experimentation and computational modeling are crucial for optimizing these new formulations to meet performance requirements.

Another key area of investigation involves the use of biodegradable binders and plasticizers. These components often constitute a significant portion of the total mass of an energetic material, and their environmental fate is crucial. Traditional binders are frequently petroleum-based and persistent in the environment. Substituting them with bio-based polymers, derived from renewable sources like starch or cellulose, offers a sustainable solution. However, integrating bio-based components can present challenges concerning mechanical properties, sensitivity, and thermal stability, which necessitate a thorough understanding of their chemical behavior and interaction with other constituents within the energetic material.

Furthermore, the design of inherently less toxic energetic materials is essential. Many conventional energetic materials contain heavy metals or other toxic elements, leading to severe environmental contamination. Designing compounds with reduced toxicity requires a deep understanding of the chemical structure-toxicity relationship, allowing researchers to tailor molecular design to minimize harmful effects. This approach demands a combination of advanced synthesis techniques and comprehensive toxicological evaluation. Utilizing computational chemistry tools to screen vast chemical spaces aids in identifying promising candidates before undertaking costly and time-consuming laboratory synthesis and testing.

The development of green energetic materials is not simply a matter of substituting one component for another. It's a holistic approach involving careful consideration of the entire lifecycle of the material, from synthesis and processing to eventual disposal or degradation. Life-cycle assessment (LCA) methodologies provide a crucial framework for evaluating the overall environmental impact of a green energetic material, considering factors such as energy consumption, resource usage, waste generation, and potential ecotoxicological effects. This comprehensive approach enables a more informed decision-making process, ensuring that the improvements made in one area do not inadvertently lead to negative consequences in another. Beyond these material-specific approaches, the development of efficient and environmentally friendly manufacturing processes is crucial. Reducing energy consumption and minimizing waste generation during production are vital aspects of achieving sustainability in the energetic materials industry. This requires innovation in manufacturing techniques, possibly utilizing greener solvents or alternative energy sources in the synthesis and formulation process. Adopting cleaner manufacturing techniques can significantly contribute to the overall environmental footprint of a green energetic material, making it a key aspect of the overall greening effort. The integration of advanced process control and automation can enhance efficiency and minimize waste during production, reducing the environmental impact even further.

The transition to green energetic materials is not merely an environmental imperative; it's also a strategic advantage. Governments and militaries increasingly prioritize the use of environmentally friendly technologies, creating a significant market demand for green energetic materials. Investing in research and development in this area not only contributes to a healthier planet but also fosters economic opportunities and strengthens national competitiveness. This transition will require collaborative efforts between researchers, manufacturers, and regulatory bodies to establish comprehensive standards and guidelines for the development, production, and deployment of green energetic materials. Only through a collaborative, multidisciplinary approach can we fully harness the potential of this field and ensure a more sustainable future for energetic materials technology.

www.ingramcontent.com/pod-product-compliance
Lightning Source LLC
LaVergne TN
LVHW022012200926
848481LV00007B/2361
9798348522360